Hotchkiss Public Library
P.O. Box 540 MAY 0 6
Hotchkiss, CO 81419

The Deepest Canyon

Stuart A. Kallen

KIDHAVEN
PRESS™

San Diego • Detroit • New York • San Francisco • Cleveland
New Haven, Conn. • Waterville, Maine • London • Munich

© 2004 by KidHaven Press. KidHaven Press is an imprint of The Gale Group, Inc., a division of Thomson Learning, Inc.

KidHaven™ and Thomson Learning™ are trademarks used herein under license.

For more information, contact
KidHaven Press
27500 Drake Rd.
Farmington Hills, MI 48331-3535
Or you can visit our Internet site at http://www.gale.com

ALL RIGHTS RESERVED.
No part of this work covered by the copyright hereon may be reproduced or used in any form or by any means—graphic, electronic, or mechanical, including photocopying, recording, taping, Web distribution or information storage retrieval systems—without the written permission of the publisher.

LIBRARY OF CONGRESS CATALOGING-IN-PUBLICATION DATA

Kallen, Stuart A., 1955–
  The world's deepest canyon / by Stuart A. Kallen.
  p. cm. — (Extreme places)
  Includes bibliographical references (p.   ).
  Summary: Describes the history, geography, animals and plants, people, exploration, and ecology of Peru's Colca Canyon, where the Colca River's elevation rises to 3,450 feet above sea level.
  ISBN 0-7377-1880-3 (hardback : alk. paper)
  1. Colca Canyon (Peru)—Juvenile literature. 2. Colca River Valley (Arequipa, Peru)—Juvenile literature. [1. Colca Canyon (Peru) 2. Colca River Valley (Arequipa, Peru)] I. Title. II. Series.
  F3451.A7K35   2004
  985'.32—dc22
                                                            2003015039

Printed in the United States of America

Contents

Chapter One
 More than Two Miles Deep   4

Chapter Two
 The Plants and Animals of
  Colca Canyon   13

Chapter Three
 People of the Canyon   23

Chapter Four
 Explorers in Colca Canyon   32

Notes   42

Glossary   43

For Further Exploration   44

Index   45

Picture Credits   47

About the Author   48

CHAPTER ONE

More than Two Miles Deep

Colca Canyon is the world's deepest canyon. At its lowest point, the canyon is more than 13,900 feet deep. That is 2,200 feet deeper than the world's next deepest canyon, which is found in Tibet. Colca Canyon is an extreme place, consisting of a narrow gorge between towering rock walls and massive piles of jagged boulders. The sheer canyon walls block out the sun for all but a few hours a day, so little grows in the canyon. In the winter, temperatures can drop well below freezing. **Avalanches** of falling snow rumble down into the canyon from the mountains high above.

The roaring Colca River runs through the canyon, smashing into rocks, shooting over waterfalls, and

---

Colca Canyon's steep, rocky walls receive only a few hours of sunlight each day.

Extreme Places • The Deepest Canyon

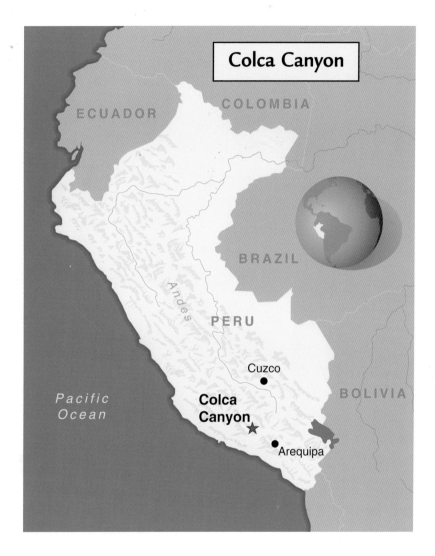

swirling into deadly whirlpools. Most of the canyon is unfit for human life, and the deepest parts have been explored by only a few people.

The Andes

Colca Canyon is located in the Andes Mountains in southern Peru. In this region the mountains are extremely high. Some reach over twenty-two thousand feet above sea level. The Andes began forming 180 million years ago when huge sheets of land, known as **plates**, slowly crashed into one another. In some places these plates were forty miles thick and forty-seven hundred miles long. As they moved over the earth's inner core of molten rock, they pushed up higher and higher to create the mountains.

The plates are still colliding. As they rub together, earthquakes can shake the ground violently under Colca Canyon. This movement, even when minor, can cause boulders as big as buildings to plummet into the canyon from high above.

When major earthquakes hit, they can be deadly. For example, in 2001 a series of earthquakes rocked the Colca Canyon region. In the small village of Maca, located in a wider, shallower part of the canyon, twenty people were killed and hundreds were injured. Almost all of the buildings in the village were destroyed.

Valley of Volcanoes

The plate movement that causes earthquakes also creates holes or cracks in the earth. These gaps allow red-hot liquid rock, called **lava**, to spew from the depths of the earth. This lava forms huge volcanoes that can rise thousands of feet above sea level.

The landscape around Colca Canyon is known as the Valley of Volcanoes. In this region there are eighty-six towering, black, snowcapped volcanoes including Mount Ampato and Mount Hualca Hualca. These formations are more than two hundred thousand years

A hiker watches as Mount Sabancaya spews black smoke and ash. The region surrounding Colca Canyon is known as the Valley of Volcanoes.

old, and most of them have not erupted for centuries. Mount Sabancaya, however, remains active. It often belches black smoke that smells like rotten eggs. Sometimes huge clouds of fiery ash spew from the open crater at its peak.

In 1990 an eruption of ash on Mount Sabancaya shot thousands of feet into the air. During the following weeks, the gray ash fell like snow from the sky. Thousands of people in the Colca Valley suffered severe breathing problems and were forced from their homes. Crops were damaged, and farm animals died.

The ash also caused snow to melt on the highest mountain peaks for the first time in centuries. This created huge rivers of thick, flowing mud that reshaped the land in Colca Canyon.

A Story in Rock

The restless rock of the high Andes also helped create Colca Canyon. This happened when the plates in the region cracked and separated, leaving a deep, dark rift of volcanic rock. Joe Kane rafted through the canyon in 1991. He described this rift as "a trench so deep and black that it [looks] like a wound in the skin of the world. . . . [I was] convinced I could follow that blackness into the earth's very guts."[1]

Over millions of years, under pressure from the heavy rock, the black lava has changed into hard granite in some places. Mixed with the granite are different

Black volcanic rock mixes with layers of lighter colored granite in the Valley of Volcanoes.

layers of younger rock. The limestone rock in the canyon was once under the ocean. It was created from the remains of bones, teeth, and seashells of ancient animals. When the plates crashed together to form the Andes, the limestone was pushed from the ocean bottom to the mountaintops. A close examination of the limestone shows the remains of these creatures still embedded in the rock.

Other rock in Colca Canyon is called sandstone. This rock, made from tightly packed sand, was once on the bottom of the ocean. Like the limestone it was shoved up high into the mountains by the movement of plates.

This jumble of rock creates a stark beauty. Bullet-shaped boulders as big as school buses perch in the sand, ready to crash over sideways at any moment. Narrow, black spires of hardened lava are capped with tons of blond sandstone. The Colca River flows past this landscape in a flurry of white-water foam.

World's Deepest Canyon

In 1981 a team of adventurers **kayaked** down the Colca River. Along the way, the men used a gauge, called an **altimeter**, to measure the altitude at the river. They found that the river elevation in the deepest part of the canyon averaged about 3,450 feet above sea level. In this area, the river flows between two massive volcanic mountains for a stretch of sixty-one miles. The peak of one of the volcanoes, Señal Yajirhua, towers 17,146 feet above sea level.

A researcher travels down the Colca river during a 1981 expedition to measure the altitude of the river.

Colca Canyon's steep cliffs and surrounding mountains make it one of the Earth's most extreme enviornments.

That means that this stretch of Colca Canyon, from riverbed to mountaintop, measures 13,969 feet. This is a distance of more than two and a half miles. No other canyon on Earth is known to be as deep as Colca Canyon.

With its massive cliffs, roaring river, and cluttered rocky terrain, Colca Canyon is one of Earth's most extreme environments. Perhaps this is why it has been given so many names over the years, including the Lost Valley and the Valley of Wonders. Because of the live volcano, it has also been called the Valley of Fire. With a two-and-one-half-mile climb straight up from the riverbed to the mountaintop, there is little doubt that the deepest part of Colca Canyon will remain a lost valley for many years to come.

CHAPTER TWO

The Plants and Animals of Colca Canyon

Hidden in a narrow gorge two miles deep, the environment of Colca Canyon is unlike any other. Within the canyon, sunlight and water are both in short supply. The Colca River is fed by melting snow, but little rain falls because the Andes block the rain clouds that blow in from the east. It rains less than six inches a year in the region, making Colca Canyon as dry as many deserts.

Tough Plants

Despite these harsh conditions, plants can be found growing in Colca Canyon. And like the canyon itself, they tend to be jagged and tough. For example, the nopal cactus, also known as the prickly pear, can grow to a height of ten to fifteen feet. This sturdy cactus can survive months without water. When it does rain, the

The nopal cactus produces red prickly pears (above) and pink flowers (right).

shallow roots of the plant quickly suck up the rainfall. This precious liquid is then stored within the paddle-shaped stems of the cactus, allowing the plant to survive through the driest periods. The long, sharp needles that jut from the paddles prevent animals from biting into the cactus for a drink.

Chapter Two • The Plants and Animals of Colca Canyon

As hard and prickly as it is, the nopal produces beautiful flowers. These blooms of pink, red, yellow, and orange form a red fruit called a prickly pear, or tuna in Spanish. With their needles removed, these prickly pears can be eaten in soup or used in various other recipes. Even the insects that live on the prickly

Cochineal bugs (shown) are used to make a natural red dye.

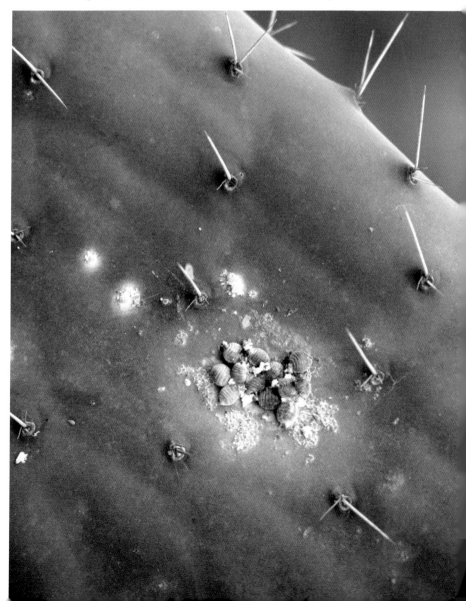

pear are useful. The cochineal bugs yield a beautiful crimson red color that is used as a natural dye.

The *yareta* is another tough plant that grows in Colca Canyon. Found in huge patches that look like moss, the slow-growing *yareta* can live for thousands of years. Able to grow on steep rock faces with very little water, patches of this rare carpetlike plant are so dense that they can be chopped apart only with a very sharp ax. Natives harvest the plant for fuel, however, because it contains an oily sap that burns with a very hot flame. The *yareta* is also used in medicine. A juice squeezed from the roots is said to relieve high-altitude sickness, a common ailment in a region where the mountain air is very thin.

Like the *yareta*, the giant *puya raimondi* is a rare and unusual plant that grows very slowly. While most plants flower every year, the *puya raimondi* takes one hundred years to produce a single huge spike forty feet high. This stem contains about twenty thousand tiny flowers that last only about three months. During this time the beautiful flowers attract hundreds of hummingbirds. After flowering, the plant dies.

Although they take a century to bloom, the *puya raimondi* are on different schedules, so that groups of flowering plants appear every three or four years. The giant puya raimondi is considered to be one of the oldest plant **species** in the world.

Extreme Birds

In a land of extremes, even the hummingbirds that flock to the giant *puya raimondi* set records. The average hummingbird in the United States has a body that measures about three and a half inches. These tiny birds weigh about an eighth of an ounce, as much as a single sheet of notebook paper. Giant hummingbirds in Colca Canyon can be twice as big and weigh ten times more. These birds, that are the size of an average starling, are the largest hummingbirds in the world.

While a seven-inch-tall hummingbird may be amazing, Colca Canyon is most famous for its condors. They weigh about twenty-five pounds and have wingspans of

Giant Andean condors (right) often soar above Cruz del Condor, or Condor Cross (pictured), four thousand feet above the canyon's lowest point.

more than ten feet. The Andean condor is the largest and heaviest flying bird in the world. (The ostrich is larger but cannot fly.) Andean condors live up to fifty years in the wild. They lay their eggs in the safety of the rock ledges high above the Colca River, and their young learn to fly when about six months old.

Andean condors are a type of vulture that eats the flesh of dead animals such as sheep, rodents, and snakes. The condors have incredibly powerful vision that allows them to spot these decaying creatures from high in the sky. Like vultures, condors have no feathers on their heads, as these would become coated with blood while feeding.

The native Inca who ruled the area more than five centuries ago believed that the condor was a sacred bird and represented the god of the skies. Today, tourists flock to a place called Cruz del Condor (Condor Cross), four thousand feet above the lowest point of Colca Canyon. Here they can see as many as thirty majestic Andean condors at one time soaring high above the canyon.

Vicuñas, Alpacas, and Llamas

Not every animal living around the world's deepest canyon is large. In fact, the ninety-pound vicuña, found throughout the Andes region, is the world's smallest camel.

Vicuñas live in small family groups on the steep mountainsides that make up the walls of Colca Canyon. Although they are only three feet tall at the

Chapter Two • The Plants and Animals of Colca Canyon

The ninety-pound vicuña is the world's smallest camel, standing only three feet tall at the shoulder.

shoulder, these animals have adapted to the extreme conditions in the area. Their long, slender legs and padded cloven hooves allow them to quickly travel across jagged, rocky terrain. Their dense tan fur keeps them warm when temperatures drop below freezing.

Vicuña fur is also highly prized for making coats and sweaters. Because the fur is so valuable, hunters have killed many vicuñas. A single pound of the fur can bring a hunter $250, a huge sum in a region wracked by poverty. Though they are nearly extinct, programs are in place to protect this tiny South American camel.

Used for their meat and for carrying heavy loads, llamas are common in Colca Canyon.

Once hunted for their fur, laws now protect guanacos.

Like vicuñas, guanacos are related to camels. Slightly larger than vicuñas, guanacos are the largest wild mammals that live in the Colca Canyon region. The animals have tan fur with white stomachs, short tails, big heads, long necks, and narrow, pointy ears up to eighteen inches long. Guanacos, too, were once hunted for their thick, warm wool but today are protected by law.

Thousands of years ago, natives chose the largest guanacos and vicuñas with the best fur and bred them. By doing this they created two new species, the alpaca and the llama. These animals are very common in the canyon area. They also serve many useful purposes. Alpaca fur is twice as strong as sheep's wool and five times warmer. Llamas are used for their meat and also to carry heavy loads on steep mountain trails.

Canyon of Extremes

Although the Colca Canyon region is cold and dry, its river, cliffs, and mountains support many hardy species of plants and animals. Some are the biggest of their species, some are the smallest. Many, such as spiders, lizards, snakes, rodents, and other creatures found in the canyon, are common in many parts of the world. But whatever their size, in this canyon of extremes, these creatures must be among the world's toughest in order to survive from day to day.

CHAPTER THREE

People of the Canyon

The Colca River is about 155 miles long. For 61 miles, it rushes though a canyon so deep, narrow, and rocky that it cannot support human life. Along other parts of the river, however, the canyon widens. Sixteen small villages are located in the wider part of Colca Canyon

Despite its remote location, people have been living in this part of Colca Canyon for at least five thousand years. The first settlers made paintings on the walls of many small caves found throughout the canyon. This rock art includes pictures of birds, vicuñas, and people hunting. Some portray the stars of the Southern Cross, a constellation that burns clearly in the night sky above Colca Canyon.

One cave, in the deepest part of Colca Canyon, harbors another ancient wonder. The Toro Muerto cemetery

Extreme Places • The Deepest Canyon

Ancient rock art (inset) and cliff-side cemeteries (shown) can be found in the many caves throughout Colca Canyon.

Children play near their house in a Colca Canyon village. People have been living in the canyon for more than five thousand years.

is located on a steep cliff face that rises straight up from the river. Here the local Indians were buried in the fetal position, in which the spine is curved, the head is bowed forward, and the arms and legs are drawn in toward the chest. Those who have studied the cemetery cannot understand how the ancients carried their dead to this remote place. Even professional climbers have difficulty descending into this part of the canyon to visit the burial ground.

The Collagua and Canaba

By A.D. 400 two other groups of people had moved into Colca Canyon. The Collagua were one group. Tribal legend states that the Collagua first came to Earth when they climbed out of a nearby snowcapped volcano called Collaguata. The tribe then walked down the mountain, chased off another tribe, and formed a village. Another group of people, the Canaba, settled to the west of the Collagua.

Although they spoke different languages, the Canaba and Collagua lived in harmony for centuries. Despite the dry, cool climate, the tribes developed a successful system of agriculture. They used **terrace** farming, in which flat, narrow stretches of ground were carved into the steep hills of the canyon like stair steps. These "staircase farms of the ancients"[2] were part of a system for managing water. Canals captured rainfall and snowmelt that would otherwise have run into the Colca River, and channeled it into the terraces. The farmers used this water to grow corn, beans, potatoes, and other food crops in the rich volcanic soil. Extra food was stored in caves, called *colcas* in the native Indian language, and this term gave Colca Canyon its name.

The terraces, which run forty miles along both sides of the wider parts of the canyon, are still used by farmers today. Because they are unable to move tractors up the steep walls, they must farm by hand, just as their ancestors did more than a one thousand years ago.

Human Sacrifice

Around the fourteenth century, the powerful Inca, who controlled much of the Andes territory, took over the villages of the Canaba and Collagua. The Inca worshipped the fire-spewing volcanoes as living gods. Mount Ampato, which towers over parts of Colca Canyon, was

Chapter Three • People of the Canyon

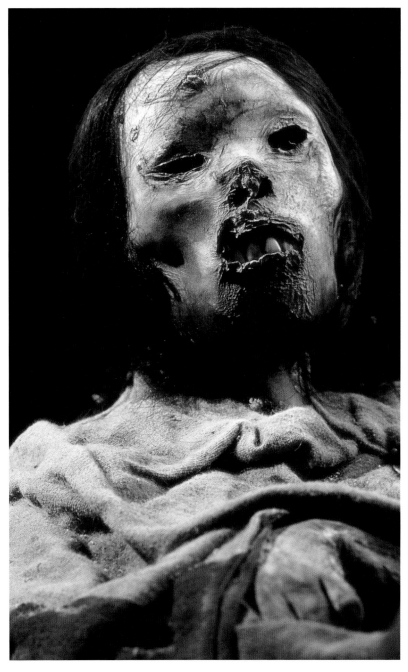

Researchers discovered Inca mummy Juanita near Mount Ampato. The Inca made human sacrifices to the volcanoes surrounding Colca Canyon.

believed to provide life-giving waters and good harvests. To ensure that this bounty continued, the Inca made human sacrifices to the god of Ampato. This often involved killing a child and leaving its remains near the summit of the mountain.

In September 1995 several researchers were climbing Mount Ampato when they discovered the body of a young Inca girl. She was tightly wrapped in cloth and had been frozen for centuries. The girl, nicknamed Juanita by the researchers, was perfectly preserved as a mummy. Juanita was dressed in beautiful clothing with colorful patterns that told researchers she was a member of Inca royalty.

Later searches revealed two other mummies on Mount Ampato. One was a young girl, and the other was a young boy whose charred body had been struck by lightning. Researchers believe that this couple was ritually sacrificed after being married.

Writers, Singers, and Musicians

When the Inca were conquered by the Spanish in the sixteenth century, a new ruler, Francisco de Toledo, came to power. Toledo wrote about the natives: "There are among them good [writers], singers, and musicians upon the flutes and [wooden pipes], and they have [abilities] for even more difficult things. [They like] feasts and banquets ... and their manners are [likeable] and but little marked by [selfishness]."[3]

The Spanish rulers had little difficulty controlling their Indian subjects. The Spanish began converting them to the

Christian religion—and forcing them to work in the local silver mines. In order to better control the natives, Toledo ordered everyone to move into twenty-four villages spread out along the sides of Colca Canyon. The towns were all similar, with Spanish-style buildings and streets that led to open plazas where churches were built. Within fifty years, however, European diseases and forced labor reduced the population by 90 percent, from about sixty thousand to six thousand people.

Ancient terraces, narrow stretches of land carved into steep hillsides, line the wider parts of Colca Canyon. Farmers still use the terraces today.

Sixteen of the isolated Spanish villages remain, and life within them has changed little in the past four hundred years. For example, the first roads to connect the villages with the outside world were not built until 1985. Before that, travel between villages was extremely difficult. The few trails that ran through the rough, black lava rock of the canyon were so rugged that a journey of five miles could take ten hours. People rode mules while hauling food, firewood, and other supplies on the backs of llamas tied together in long trains of six to twelve animals.

Farmers sell fruit near a sixteenth-century Spanish church in a Colca Canyon village.

Hikers walk along a dusty village road above Colca Canyon. Earthquakes have damaged many of the roads in the region.

Although roads built in recent decades have made it easier to travel between villages, many of these routes are little more than dusty, bumpy gravel paths. Many of the roads have been damaged by earthquakes, which are common in this region.

The people of Colca Canyon have dealt with volcanoes, earthquakes, and a harsh climate for thousands of years. Living in small villages high in the Andes, many have had little contact with the outside world. On the edge of the world's deepest canyon, life continues much as it did when the Collagua and Canaba lived in the valley more than one thousand years ago.

CHAPTER FOUR

Explorers in Colca Canyon

Although Colca Canyon was shown on nineteenth-century maps, the region was all but unknown to outsiders. That changed in 1930 when U.S. Navy lieutenant George R. Johnson and Robert Shippee flew their small, single-engine plane into Colca Canyon. Before the two men landed their plane, however, they explored the region on mules to find a flat piece of ground to use as a landing field. On this part of their journey, they rode for days on narrow, twisting trails cut into the mountainside. Shippee described his journey through the canyon:

> [Every] few minutes we had to [get off the backs of our mules] and edge our way along yard-wide slippery trails that seemed to be hung in mid air

A team of explorers rafts through a narrow gorge at the bottom of Colca Canyon.

hundreds of feet above the river.... Travel on the [canyon] floor was impossible, since the deep gorge of the river [snaked] from side to side. It was easy to see why there was little [travel between villages] in the Colca valley.[4]

After days of rugged exploration, Shippee finally found a suitable site for a landing field. Shippee hired 165 native workers to build an airstrip and paid them all a total of six dollars. Several days later, the silence of Colca Canyon was broken by the loud roar of a gaso-

Two adventurers drop into the Colca River in watertight boats called kayaks.

line engine. Johnson and Shippee landed their plane, stayed in the canyon for a day, and then flew home. They never returned.

The experiences of Johnson and Shippee were published in *National Geographic* magazine in 1934. This was the first time in history that the outside world was able read about the people of Colca Canyon. The article also contained nearly two dozen photographs, many from the air, that showed the extreme **geography** of the canyon and the rugged volcanoes in the region.

The Polish Kayak Team

Nearly fifty years later another group of adventurers arrived at Colca Canyon. This group included six Polish adventurers. They arrived in Chivay, the largest village in the canyon region, carrying bright yellow kayaks on the back of an old truck. The men told the surprised natives that they planned to be the first people to float down the river through the deepest part of the canyon.

The team, led by Andrew Pietowski, had begun its journey in Mexico in 1979. At that time, the Polish kayakers hoped to spend six months running as many rivers as possible in the mountains of Central and South America. They called themselves Canoandes (for "canoe the Andes"). Their odyssey stretched into two years. During this time the group paddled twenty-three rivers from Mexico to the southern tip of Argentina. Thirteen of these rivers had never been run. Out of these thirteen, the Colca River, pouring through the world's deepest canyon, was the most violent and dangerous.

Extreme Places • The Deepest Canyon

Giant boulders, waterfalls, and sudden drops in elevation presented many obstacles for the kayak team. In many places, the Canoandes group was forced to leave the water and **portage**, or carry their boats and supplies overland. These portages were brutal in places where avalanches blocked the way and rock walls jutted two miles straight up from the riverbed. In such areas, the team used ropes and rock-climbing equipment to move their expedition downstream. Presented with these hazards, the men of Canoandes wanted to quit. As Chmielinski says, "we were *scared*. Two times we tried to climb out, but we could not. We see all these landslides around us and we think . . . we are *crazy* to be here."[5]

A kayaker flies headfirst over a Colca River waterfall. The river's many obstacles make it extremely dangerous.

Kayakers from the Polish team paddle through the rushing rapids of the Colca River in 1991.

Forced to complete their journey, the team ran into other problems as well. The kayakers got sick and ran out of food. At times, men were almost killed by rocks hurtling down from the mountains above. Despite these dangers, the members of Canoandes became the first people to ever successfully run the world's deepest canyon.

In 1983 the Polish team ran the Colca once again. This time they brought Alvaro "Cholo" Ibáñez, who became the first Peruvian to run the river. Ibáñez returned to the dangerous river in 1985 with his own team of four, three Peruvians and a woman from Belgium. But tragedy struck. Heavy rains and snow had flooded the Colca, and the team's raft overturned less than one minute into the journey. Three men survived, but Ibáñez's body was never found.

Returning to the River

Fully aware of the hazards he faced, Chmielinski decided to run the river once again in 1991. This time he

Members of the Polish kayak team use ropes to climb up a rocky canyon wall in 1991.

put together a team of twelve, three of whom were veterans of the first Colca run. The team brought two four-man rafts and four kayaks.

Because of earthquakes and avalanches, the course of the river had changed and had become even more hazardous since his first run. To avoid disaster, the team followed a routine. The four light kayaks, which are easily steered, went down the river first. The kayakers then climbed up the rocky banks with ropes attached to the heavy rafts, which were laden with supplies and difficult to maneuver. If a rafter fell into the water, the kayakers would attempt to rescue him. Great care had to be taken at all times because a person who was injured would have no way out of the deep canyon.

"We Are Dead Meat!"

Although constant caution was needed, it was very difficult to sustain. Crew members were tired all the time, working from sunup to sundown running the violent, frothy rapids. They were also wet and cold all day long, with water constantly rushing into their boats.

Dangers increased in places where the fast river was shallow. The rafts often snagged on sharp rocks. The crew would then fight desperately to free them before they overturned. One member of the team, Joe Kane, describes a situation that almost turned deadly:

> [Our] raft floor just exploded. No one had seen the razor-tipped rock just below the surface....

Extreme Places • The Deepest Canyon

There was a sound like a rocket firing, and we went [flying]. I [was] . . . suspended right over the hole, inches from the howling white [water] that wanted to eat me alive, and somewhere I could hear Piotr screaming . . . 'We are dead meat!'[6]

The crew survived and patched the raft later in the day. By the fourth day, however, almost half the team had the flu and food was running low. Although they were working to the point of exhaustion, each day lunch was made up of only two pieces of salami, some stale bread, and a few peanuts. At night they slept in ab-

Chmielinski's crew cling to the raft as they are swept over a waterfall. Colca Canyon remains one of the world's most extreme places.

solute darkness as two miles of upright rock closed in overhead.

For eleven days, the team floated down the raging river, hungry, wet, and tired. Sometimes they were in complete control, other times they simply flew over twenty-foot waterfalls and hoped for the best. At one point, Kane almost drowned in swirling rapids. "It was like one of those stomach-churning carnival rides, but with no guarantee it would stop. I whirled and spun. I went limp; no use fighting it. I waited. And I waited. And I waited, helpless, while the river beat me at will. Then, for no . . . reason: Air!"[7]

A Place of Wonder

In the years since Chmielinski's and Kane's journey, few others have tried to run the deepest stretch of the Colca River. In this extreme rift cleaved into the earth, the river, avalanches, volcanoes, and earthquakes remain in control. And as the forces of nature continue to shape it year after year, the world's deepest canyon will remain what is has always been: a place of mystery, wonder, and deadly danger.

Notes

Chapter 1: More than Two Miles Deep
1. Joe Kane, "Roaring Through the Earth's Deepest Canyon," *National Geographic,* January 1993, p. 122.

Chapter 3: People of the Canyon
2. Quoted in Patti Moore, "Colca's Elusive Waters," *Americas,* January/February, 1990, p. 40.

3. Quoted in Robert Shippee, "A Forgotten Valley of Peru," *National Geographic,* January 1934, p. 116.

Chapter 4: Explorers in Colca Canyon
4. Shippee, "A Forgotten Valley of Peru," p. 122.

5. Quoted in Kane, "Roaring Through the Earth's Deepest Canyon," p. 130.

6. Kane, "Roaring Through the Earth's Deepest Canyon," p. 131.

7. Kane, "Roaring Through the Earth's Deepest Canyon," pp. 135–36.

Glossary

altimeter: An instrument used to determine elevation that works by sensing air-pressure changes that go along with changes in altitude.

avalanche: A fall or slide of a large mass of snow, rock, or mud down a mountainside.

geography: The physical features of the earth's surface, such as hillsides, mountains, lakes, and so on.

kayak: A watertight canoe with a single or double opening in the center and propelled by a person using a double-bladed paddle. Originally made from animal skins by Eskimos, kayaks today are most often made from plastic.

lava: Molten rock that reaches the earth's surface through a volcano or fissure.

plates: Sections of the earth's crust that are in constant motion.

portage: To carry boats and supplies across land and around obstacles that lie in the navigation path of a waterway.

species: A category or type of plant or animal.

terrace: A flattened bank of earth used for farming carved into a hillside and held in place with stone walls.

For Further Exploration

Books

Boy Scouts of America, *Whitewater.* Irving, TX: Boy Scouts of America, 1998. Information about whitewater canoeing, kayaks, and kayaking.

Larry Dane Brimner, *Valleys and Canyons.* New York: Childrens Press, 2000. Describes different kinds of valleys and canyons and how they are formed.

Bobbie Kalman, *Peru: The Land.* New York: Crabtree, 2003. A book about the diverse geography, plants, and wildlife of Peru, from coastal deserts to the Andes.

Johan Reinhard, *Discovering the Inca Ice Maiden: My Adventures on Ampato.* Washington, DC: National Geographic Society, 1998. A first-person account of the 1995 discovery of the more-than-five-centuries-old Peruvian ice mummy on Mount Ampato.

Internet Source

Bonnie Hamre, "Colca Canyon, Peru," *South America for Visitors,* 2003. http://gosouthamerica.about.com/library/weekly/aa010401a.htm.

Index

alpacas, 22
altimeters, 11
Andean condors, 18
Andes
 formation of, 7, 11
 heights of, 6–7
 rainfall and, 13
animals, 14, 18, 20–22
ash, 8–9

birds, 16–18
burials, 25

cactus, 13–16
camels, 18, 20–21
Canaba, 26
canals, 26
Canoandes, 35–37
caves
 burials in, 25
 food stored in, 26
Chivay (Peru), 35
Chmielinski, Piotr, 9, 35–37, 38–41
Christianity, 28–29
Colca River
 course of, 39
 length of, 23
 movement of, 4, 6
 water for, 13
colcas (caves), 26
Collagua, 25–26
Collaguata (volcano), 26
condors, 17–18

Cruz del Condor (Condor Cross), 18

depth, 4, 11–12
diseases, 29

earthquakes, 7
endangered animals, 20

farming, 26
flowers, 15, 16
fuel, 16
fur, 20, 21–22

giant *puya raimondi*, 16
granite, 9, 11
guanacos, 21

human sacrifice, 26, 28
hummingbirds, 16–17

Ibáñez, Alvaro "Cholo," 37–38
Inca, 18, 26, 28
Indians, first, 23, 25
insects, 15–16, 22

Johnson, George R., 32, 34–35
Juanita, 28

Kane, Joe, 9, 39–41

lava, 7, 9
limestone, 11

llamas, 22, 30
Lost Valley, 12

Maca (Peru), 7
medicine, 16
Mount Ampato, 7–8, 26, 28
Mount Hualca Hualca, 7–8
Mount Sabancaya, 8
mummies, 28

names, 7, 12
National Geographic (magazine), 35
nopal cactus, 13–16

Peru, 6, 7, 35
plants, 13–16
plates, 7, 9
population, 29
prickly pear cactus, 13–16

rainfall, 13, 26
religion, of Inca, 26, 28
roads, 30–31
rock art, 23
rocks, 9, 11
rodents, 22

sandstone, 11
Señal Yajirhua, 11–12

Shippee, Robert, 32, 34–35
snakes, 22
Spaniards, 28–30
spiders, 22
staircase farms of the ancients, 26
sunlight, 4

temperatures, 4
terrace farming, 26
Toledo, Francisco de, 28–29
Toro Muerto cemetery, 23, 25
tuna (fruit), 15

Valley of Fire, 12
Valley of Volcanoes, 7–9
Valley of Wonders, 12
vicuñas, 18, 20
villages, Spanish, 30
volcanoes
    activity of, 8–9
    formation of, 7–8
    height of, 11–12
    Indians and, 25–26, 28
    number of, 7
vultures, 18

wool, 21, 22

yareta, 16

Picture Credits

Cover: © Grant Dixon/Lonely Planet Images
© Stephen L. Alvarez/National Geographic, 8, 27
© Jeremy Blough, 12, 17, 31
© Jacek Boguki, 10, 24 (inset), 33, 34, 36, 37, 38, 40
COREL Corporation, 14(both), 20
© Mark Daffey/Lonely Planet Images, 5
© Ecoscene/CORBIS, 15
© Jason Edwards/Lonely Planet Images, 21
Chris Jouan, 6
© Charles & Josette Lenars/CORBIS, 9
© Mike Maass, 24, 25
© Buddy Mays/CORBIS, 17(inset)
© Reuters NewMedia, Inc./CORBIS, 29
© Paul Souder/Getty Images, 19
© Wes Walker/Lonely Planet Images, 30

About the Author

Stuart A. Kallen is the author of more than 150 nonfiction books for children and young adults. He has written on topics ranging from the theory of relativity to the history of rock and roll. In addition, Mr. Kallen has written award-winning children's videos and television scripts. In his spare time, Stuart A. Kallen is a singer/songwriter/guitarist in San Diego, California.